D0742158

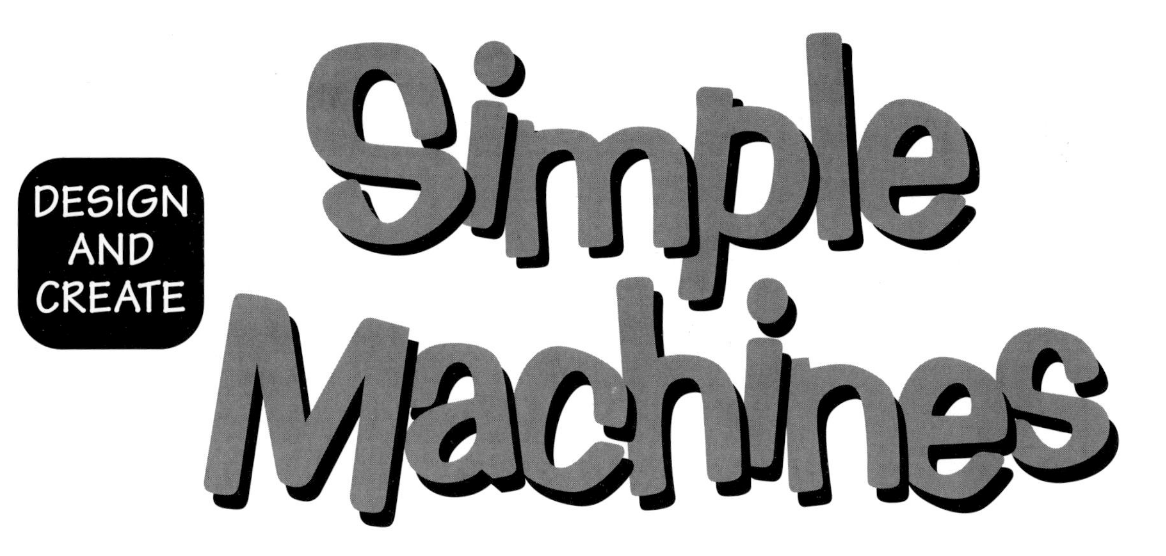

Fran Whittle
Sarah Lawrence

Austin, Texas

Published by Raintree Steck-Vaughn Publishers,
an imprint of Steck-Vaughn Company

Library of Congress Cataloging-in-Publication Data
Whittle, Fran.
Simple machines / Fran Whittle, Sarah Lawrence.
p. cm.—(Design and create)
Includes bibliographical references and index.
Summary: Twelve hands-on activities present smple machines at work
ISBN 0-8172-4889-7
1. Simple machines—Juvenile literature.
[1.Simple machines—Experiments. 2. Experiments.]
I. Lawrence, Sarah. II. Title. III. Series: Design and create.
TJ147.W485 1997
621.8'11—dc21 97-20319

Printed in Italy. Bound in the United States.
1 2 3 4 5 6 7 8 9 0 02 01 00 99 98

Commissioned photography by Zul Mukhida

CONTENTS

INTRODUCTION

Machines must have some energy put into them to make them work. The simplest machines use human or animal energy, but others use water, wind, tension, gravity, electricity, oil, and gas.

Many mechanisms are very simple. More complicated ones are often made up of a lot of simple parts. Some mechanisms can do several jobs at once, work at different speeds, and are made from many materials.

Throughout history, people have made machines because they wanted to do things that needed extra strength or speed, such as plowing a field or traveling in a cart pulled by a horse. Machines can also do work that involves repetitive actions: for example, putting food in packages.

Machines are getting more complicated all the time, and some have computers built into them. This machine is a type of robot. It is building cars.

Musical instruments have mechanisms that are worked by people to make sounds. Bagpipes have air blown into them. The air is then pumped out by squeezing the bag.

Making good machines came through trial and error, sometimes accidentally, and by knowing how the human body worked.

A machine needs to be right for the job to be carried out. The design must use materials that are strong enough and must consider environmental effects such as fumes from gasoline engines.

Many toys have moving parts. These toys are all more than 100 years old. Some are just pushed along. Others are wound up with a key, like an old clock.

WAVING ARMS

YOU WILL NEED

- Cardboard box, about 10 in. (25 cm) square
- Three strips of thick cardboard, about 2–3 in. (6 cm) wide and 14 in. (35 cm) long
- Poster board for face and hands
- Large metal paper fastener
- Hole punch
- Glue stick
- Pencil
- Ruler and scissors
- Paint and paintbrushes

This is a machine that uses levers. It has three levers that are all joined. Pull and push one lever to make the others work in different directions.

1 Using the pencil and ruler, measure the center line of the box, up both sides, and over the top.

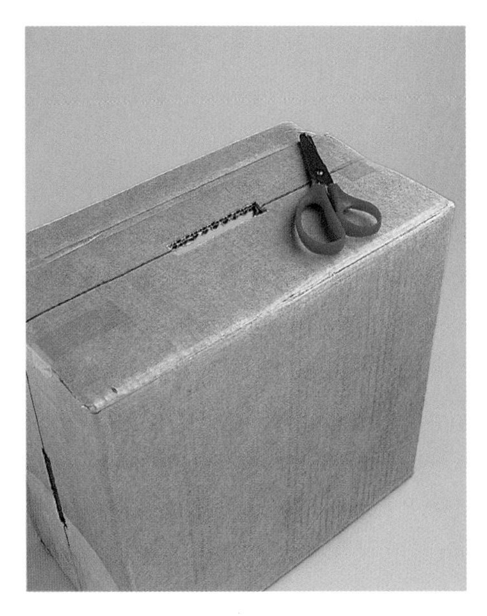

2 In the center of each line, mark a slot just big enough to take the cardboard strips. Each slot should be about 2.5 in. (6.5 cm) long and .5 in. (1 cm) wide.

A seesaw is a very large lever with the pivot in the middle. Pushing down on one end makes the other end rise. This simple mechanism has been used for different machines since ancient times.

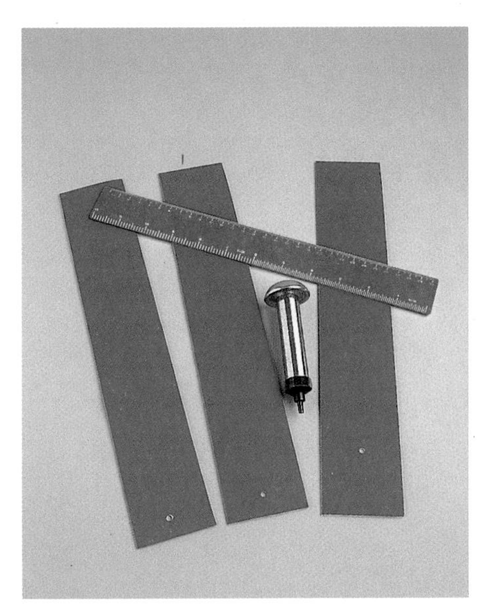

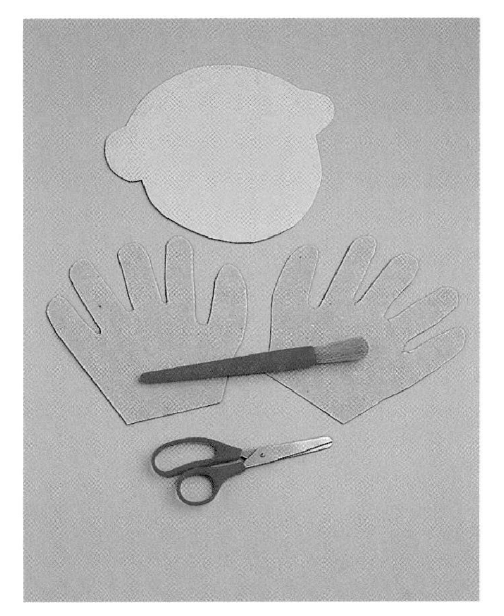

3 Make small holes at the ends of the cardboard strips. On two strips the holes should be about .75 in. (2 cm) from one end. On the other strip, the hole should be about 2 in. (5 cm) from the end.

4 Draw a face and hands on the poster board. Cut them out, and paint them. Also decorate the box.

5 Glue the hands to the two shorter strips of cardboard, at the ends opposite the holes. Glue the face to the longer strip.

6 Bend the longer strip about 1 in. (3 cm) from the end with the hole. Slide the ends of the strips into the slots in the box. Fasten them together loosely with the paper fastener.

7 Pulling a side strip in and out will make the top strip wave back and forth. Pushing the bent handle of the long strip up and down will make the side strips wave up and down.

MYSTERIOUS ROLLERS

Impress your friends with a mystery machine that has been a toy for many years.

YOU WILL NEED

- Tall cardboard box
- Two wide cardboard tubes at least 4 in. (10 cm) longer than the width of the box
- Used gift-wrapping paper
- Glue stick and tape
- Pencil
- Ruler
- Scissors
- Paint and paintbrushes (optional)

1 If the box has flaps, close them and tape them shut. On two opposite sides of the box, mark with a pencil and then cut out most of the sides. Leave some cardboard behind at the top and the bottom, to make the box stronger.

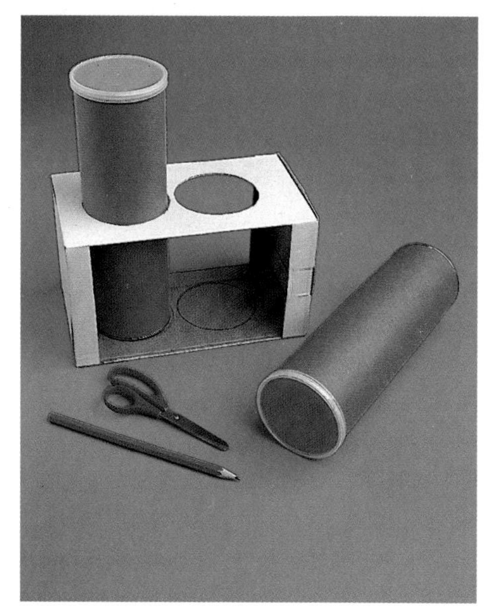

2 On one side of the box, put the ends of the tubes about .5 in. (1 cm) apart. Draw around them and cut out the holes. Push the tubes through and draw around them on the other side. Cut out the circles. Push the tubes through the box.

Rollers are used in many different machines. These women in Bangladesh are using rollers to make paper.

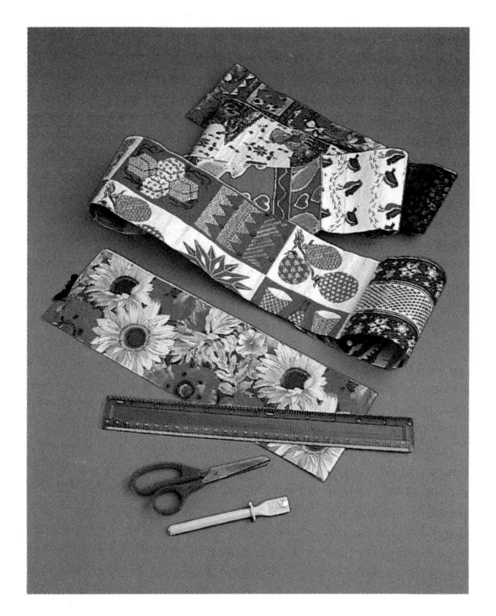

3 Cut the wrapping paper into strips wide enough to fit on the rollers inside the box. Glue them together to make a piece about 5 ft. (1.5 m) long that is patterned on both sides.

4 Tape one end of the paper strip to the top roller, and wind all the paper onto it tightly. Thread the other end between the rollers and tape the end to the lower roller on the other side, so that it makes an S shape from the side. Wind a little paper back onto the lower roller.

5 Draw and cut out one shape in two different colors. Wind one shape into the rolled paper. Keep turning, then turn the other roller and wind the second shape into the paper on the other side. It will seem to disappear, and as you keep winding, the first shape will appear again.

6 Decorate the box to match the paper. It helps if you mark or decorate the box to remember which side is which.

NOW TRY THIS

Make some shapes that look as if they change into each other in the rollers: for example, a cat into a dog, a superhero into a mouse, or a tadpole into a frog.

FLYING MESSENGER

YOU WILL NEED

- Stiff wire about 18 in. (45 cm) long
- Plastic model propeller, about 6 in. (15 cm) long
- Large rubber band
- Two plastic beads, about .5 in. (1 cm) diameter
- Poster board
- Colored paper
- Very smooth cord or fishing line
- Tape and scissors
- Pliers
- Paper clip
- Felt-tip pens

The energy stored in a twisted rubber band can turn a propeller. You can fly the messenger along a line stretched between two walls or two pieces of furniture. Attach a message to surprise your friends or family.

You will need to ask an adult to help you bend the wire.

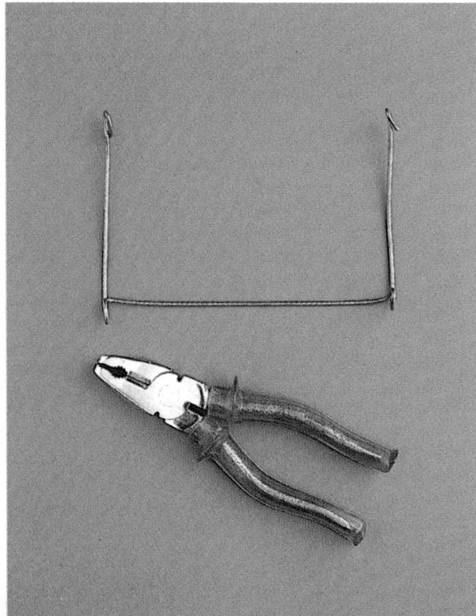

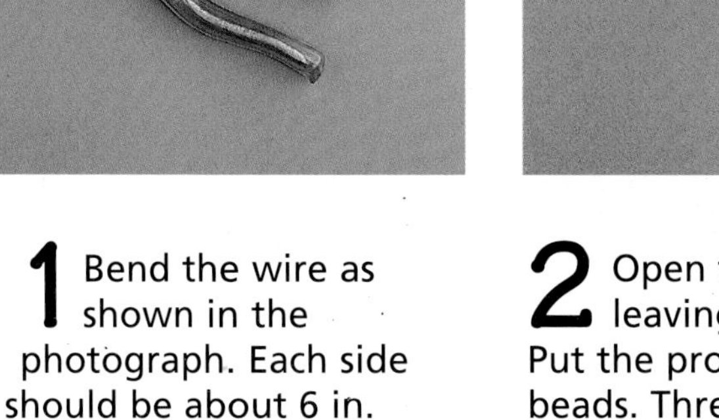

1 Bend the wire as shown in the photograph. Each side should be about 6 in. (15 cm) long. Make a closed loop at each end of the wire.

2 Open the paper clip leaving a hook at one end. Put the propeller on it, then the beads. Thread the straight end of the paper clip through a wire loop and make it into a hook again.

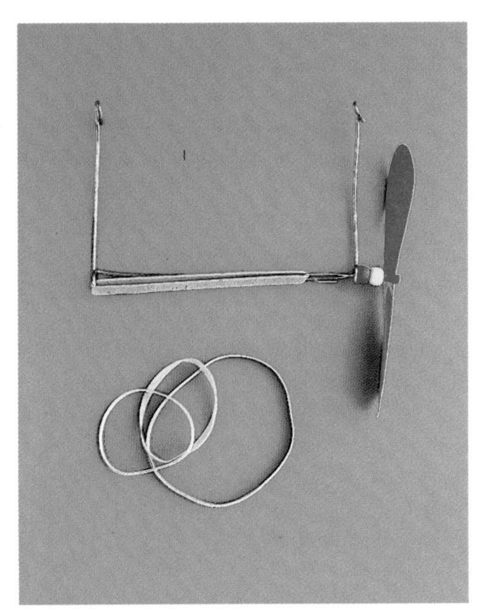

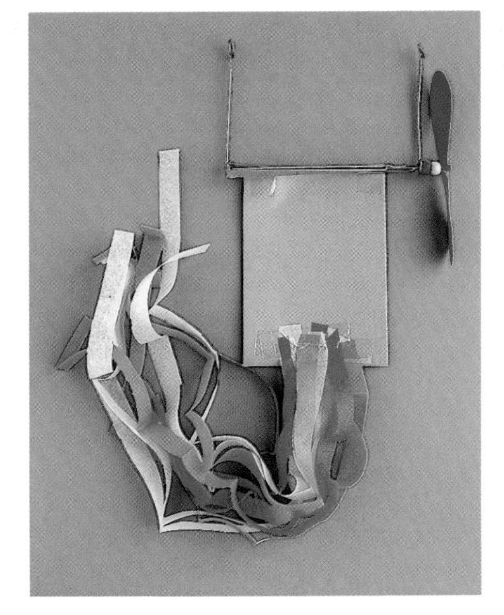

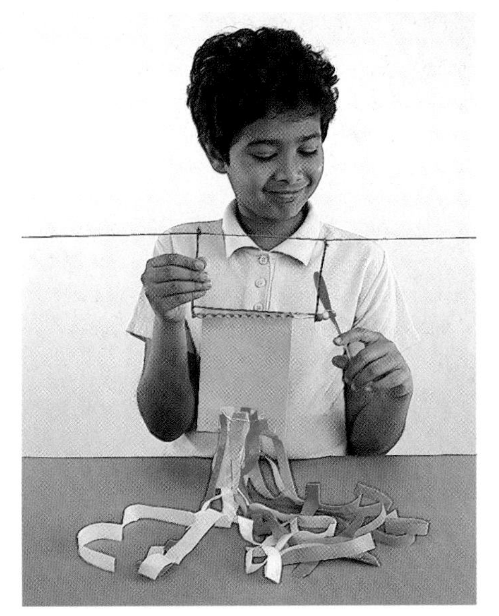

3 Put the rubber band on the other wire loop and stretch it so that it goes over the hook of the paper clip. Make sure the band is stretched so that it stays on.

4 Write a message on the poster board. Decorate the lower edge with streamers made of colored paper. Fasten the top of the poster board to the wire frame with tape.

5 Thread the line through the top loops of the flyer. Tie the line in position. Put your machine in the middle of the line. Wind the propeller at least 50 times and let go.

NOW TRY THIS

Fly a message to a friend next door. Change the size of the wire frame and the propeller and see what happens.

Old clocks get their energy from springs made of metal. The springs are wound up until they are tight. As they unwind, they let go of energy in the same way as the wound rubber band that is used on this page.

WEIGH-YOUR-SAVINGS BANK

Make a money bank that uses gravity to show how much you have saved. The more money you add, the farther the pointer will go down the box.

YOU WILL NEED

- Cardboard box with lid, such as a shoe box
- Medium-size paper clip
- Small plastic container
- Thin wire
- Two small rubber bands
- Poster board
- Tape
- Stapler
- Pencil
- Ruler
- Scissors
- Felt-tip pens

1 In one end of the box, make two holes in the center. Cut a slot for coins between the two holes and the bottom edge of the box. Make sure this slot is also in the center.

2 Unbend a paper clip and hook one end through the two holes. Let the other end hang down into the box.

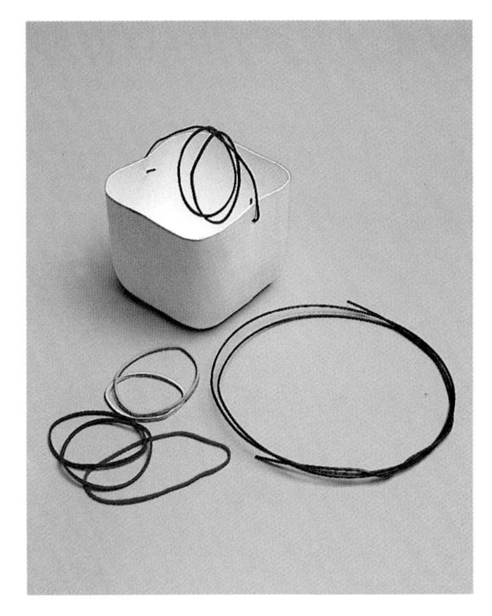

3 Make two small holes on either side of the plastic container. Make a wire handle. Put the two rubber bands on the handle.

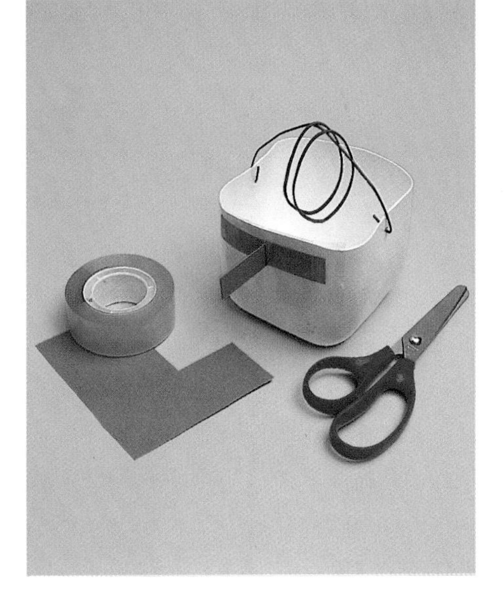

4 Cut two pieces of cardboard 2 in. x .5 in. (5 cm x 1 cm). Staple them together in the middle. Open out the ends on one side. Tape them firmly to the side of the container.

Weighing is one way of figuring out how much there is of something. Food is often weighed so that people know how much to pay for it. This man is weighing and selling bananas in Greece.

5 Cut a slit about .25 in. (6 mm) wide down the center of the bottom of the box. Start it about 3 in. (8 cm) from one end of the box. Hang the container on the hook of the paper clip.

6 Poke the cardboard strip stuck to the container through the slit. Make an arrow and stick it to the strip poking through the slit. Draw the weighing scale next to the slit. Put the lid on the box and hold it on with rubber bands. Now start saving.

NOW TRY THIS

Design and make a portion-weighing machine for food. It could be for something such as rice, breakfast cereal, or even dry pet food.

DRUM MACHINE

This machine makes a steady rhythm. Turning the handle changes the movement from around and around to up and down. This is the same mechanism as the one used in sewing machines.

This Chinese woman is using a sewing machine that is driven by her feet. The wheel near her feet is joined with a drive belt to the wheel on the machine. The machine changes the movement of the wheel to a movement that makes a sewing needle go up and down.

YOU WILL NEED

- Medium-size cardboard carton, about 10 x 6 x 6 in. (25 x 15 x 15 cm)
- Two strips of stiff cardboard, about 3 in. (7.5 cm) wide and 18 in. (45 cm) long
- Piece of straight, stiff wire, about 12 in. (30 cm) longer than the cardboard carton
- Two large paper clips
- Four thread spools
- String or thick thread
- Large rubber band
- Tape
- Pliers
- Scissors

1 Make a small hole in one strip of cardboard, about 1 in. (2.5 cm) from the end. Cut a slot in the end of the other piece, about 1 in. (2.5 cm) long.

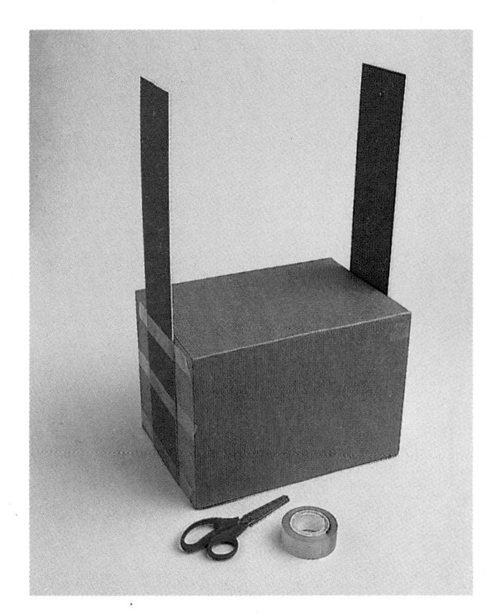

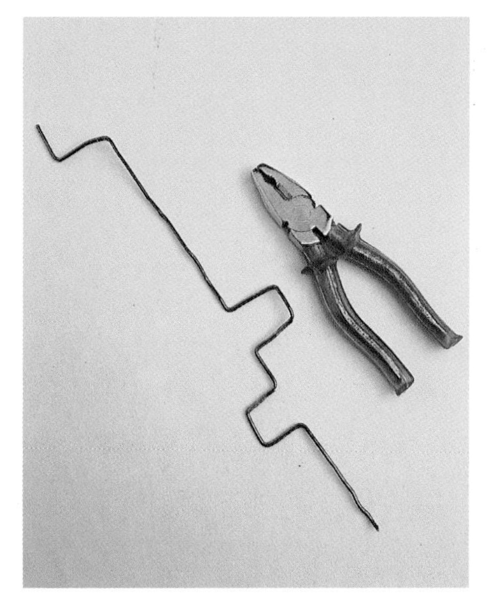

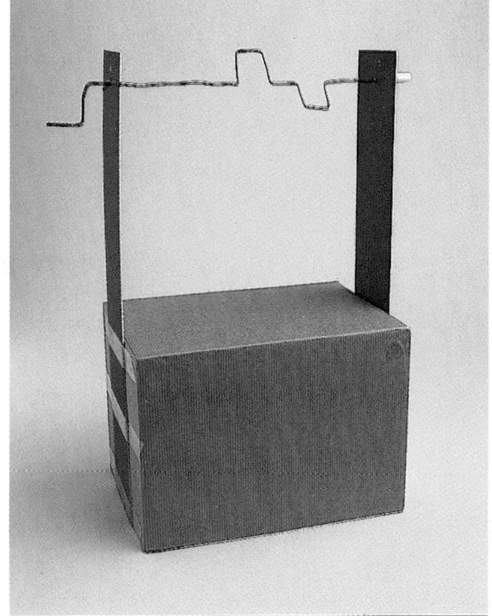

2 Turn the box so that a smooth side faces upward. Stick the strips to the sides of the box with tape.

3 Bend the wire into the shape shown in the photograph. The shapes in the middle are called cranks. They must fit between the two cardboard strips. The piece of wire is called a crankshaft.

4 Put the end of the crankshaft into the hole and the handle end into the slot. Put a rubber band over the slot to hold the crankshaft down. Wind tape around the shaft's end to hold it in place.

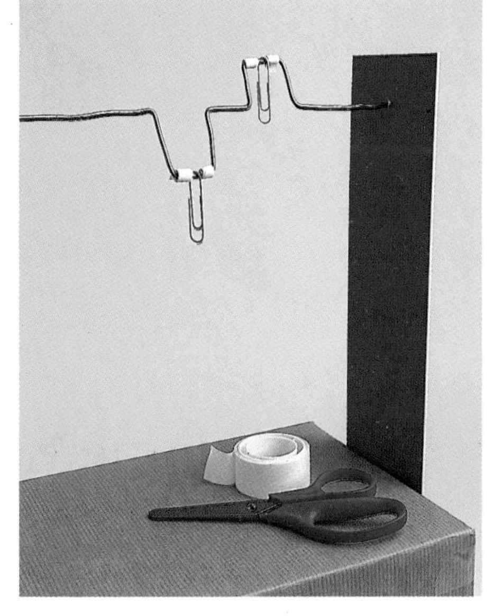

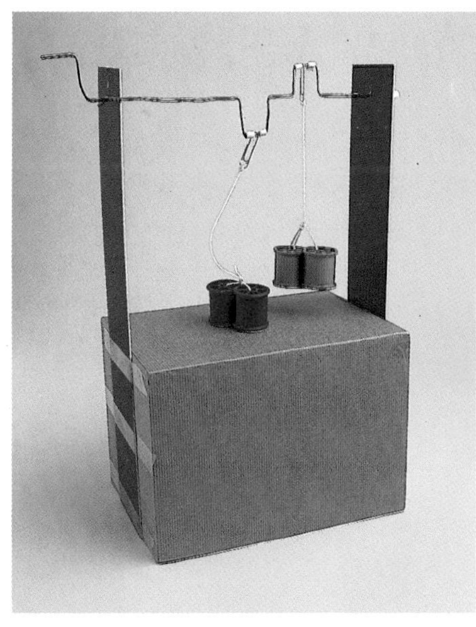

NOW TRY THIS

- **Try out different strikers and surfaces, such as upside-down yogurt containers striking a cookie sheet or wooden beads striking wood.**
- **Make a machine with a larger box and a longer crankshaft with more cranks and strikers.**

5 Put paper clips on the bends in the crankshaft and wrap tape on either side of each crank to keep the paper clips loosely in place.

6 Tie two spools onto a piece of string. Tie the other end to a paper clip. Turn the handle to check that the reels strike the box, and change the length of the thread if necessary. Repeat with the other spools and paper clip.

MOUSE-A-PULT

Here is a machine that lets you play a fun game with a cat. The mouse is catapulted away from the box. You can then "rescue" the mouse by winding it back into its hole.

YOU WILL NEED

- Medium-size, stiff cardboard box, about 8 x 8 x 4 in. (20 x 20 x 10 cm)
- Thin plastic ruler
- Two pieces of wooden dowel, about .25 in. (6 mm) diameter and about 4 in. (10 cm) longer than the width of the box
- Cardboard tube, about 1.5 in. (4 cm) diameter
- Three large rubber bands
- Small "mouse" made from paper, fabric, or an old sock
- A 5-ft. (1.5-m) piece of string
- Glue and scissors
- Hole punch
- Hand drill and drill bit

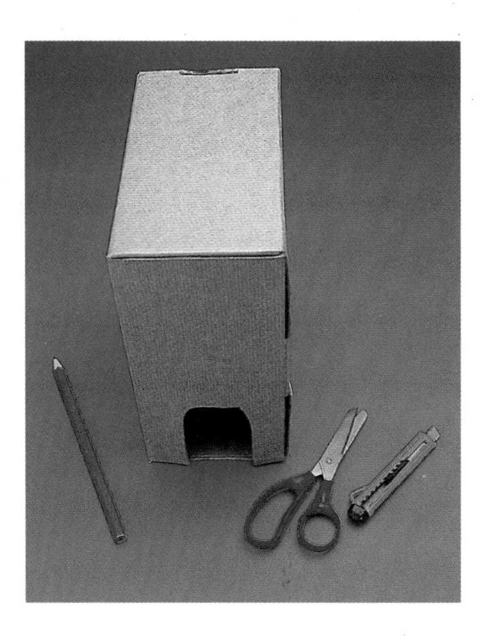

1 Cut a mouse hole in the box. It is easiest to do this on the bottom of the side that opens. Cut a slot for the ruler on the top edge of the opposite side.

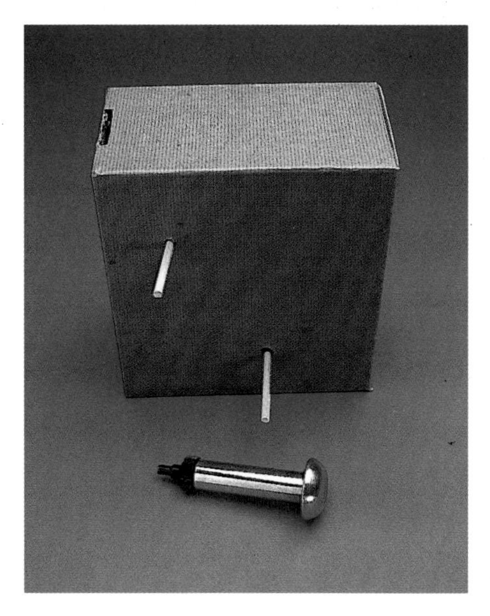

2 Punch holes in the sides of the box for the two dowels. Make sure the holes are opposite each other. Push the dowels through.

Catapults were used in fighting by the ancient Greeks and Romans and in the Middle Ages. See the catapult throwing rocks in the bottom right corner of this old painting.

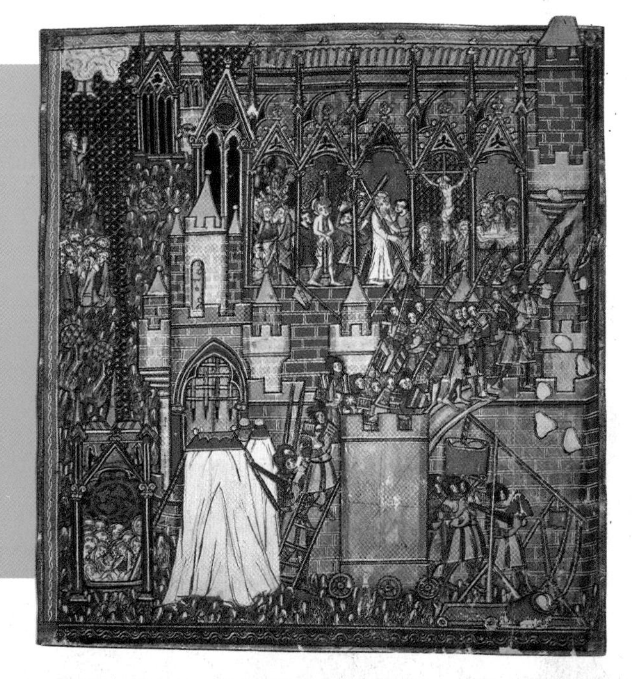

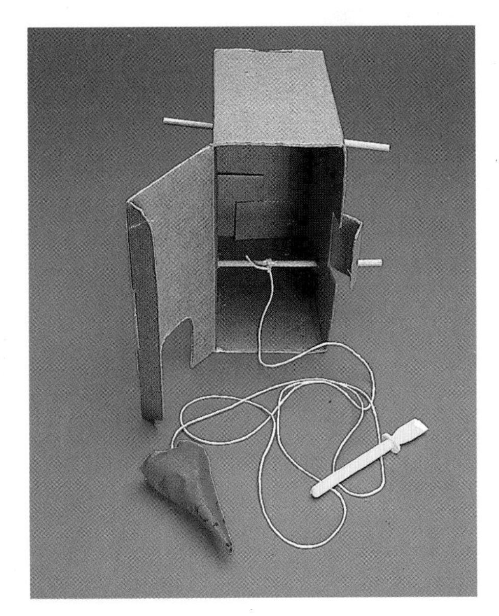

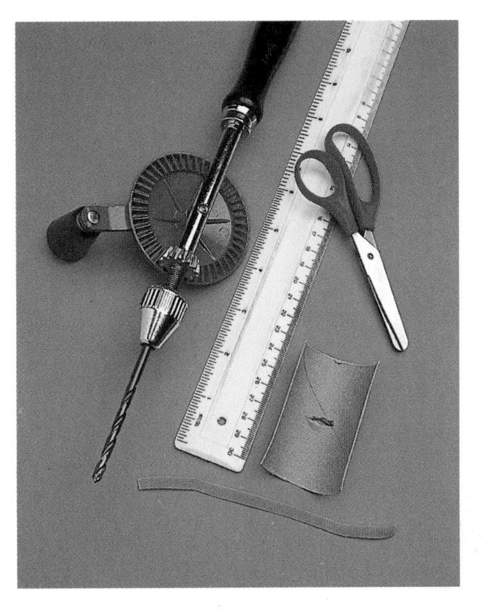

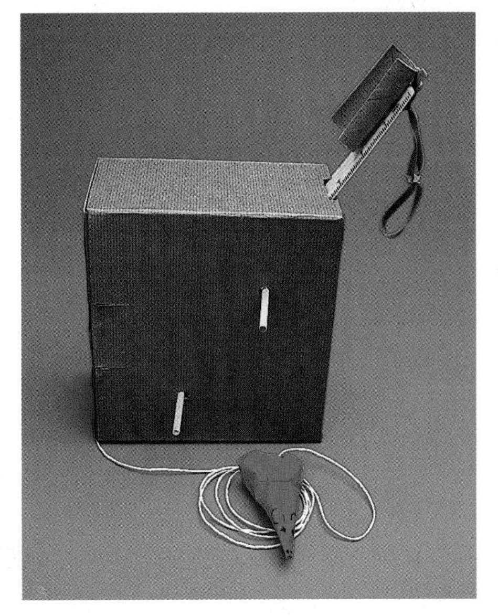

3 Tie one end of the string to the mouse. Tie the other end to the middle of the dowel near the mouse hole. Add a little glue to hold the string on the dowel.

4 Ask an adult to help you drill a small hole in the ruler, about 1 in. (2.5 cm) from one end. Cut some cardboard tube to form a cradle for the mouse. Tie the cradle onto the ruler, through the hole, with a rubber band.

5 Put the other end of the ruler into the slot on the top of the box. Make sure it goes down behind the two dowels. Tie on two more rubber bands at the top.

6 Put the mouse-a-pult at least 6 ft. (2 m) away from the cat. Put the mouse in the cradle. Make sure the string is hanging clear of the dowels. Put one hand on the box to hold it and pull down on the rubber bands. Let go and fire the mouse toward the cat. Slowly or quickly turn the bottom dowel until the mouse disappears into the hole, chased by the cat.

NOW TRY THIS

Find out what is the longest piece of string you can use. Find out if the weight of the mouse makes any difference.

MINI SPREADER

YOU WILL NEED

- Corrugated plastic (or cardboard)
- Top of a plastic 35mm film canister
- Medium-size paper clip
- Old plastic pen top or .75 in. (2 cm) length of plastic tubing
- Three small metal paper fasteners
- Small foil pie pan or plastic lid, about 3–4 in. (8–10 cm) diameter
- Masking tape
- Pliers and bradawl
- Pencil
- Ruler
- Scissors

Be careful when using a bradawl to make holes.

- **Hold the object in a vise if possible.**
- **Ask an adult to show you what to do.**

Use this mini machine for spreading sugar on your breakfast cereal or cinnamon on a slice of buttered toast.

You may need to ask an adult to help you make holes in parts of this machine. Start with really small holes: you can always make the holes larger, but not smaller.

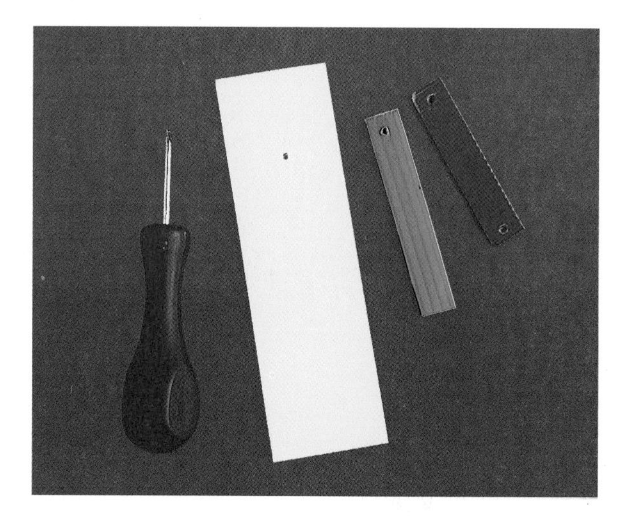

1 Cut a piece of corrugated plastic 8 x 2 in. (20 x 5 cm). Make a hole about 1.5 in. (4 cm) from one end. Cut another piece about 3 x .5 in. (8 x 1 cm), and make a small hole in one end. Cut a third piece about 2.5 x .5 in. (6 x 1 cm) and make small holes in both ends.

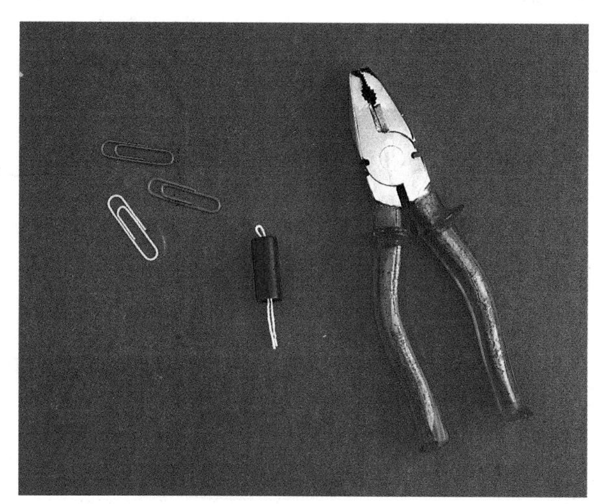

2 Using the pliers, open out the paper clip, then bend it in half. Make a hole in the top of the pen top with the bradawl, and push the ends of the wire through, leaving a little loop sticking out at the top.

Wipers on windshields work in the same way as the Mini Spreader. The wipers are kept in check so that they just cover the windshield, rather than going all the way around.

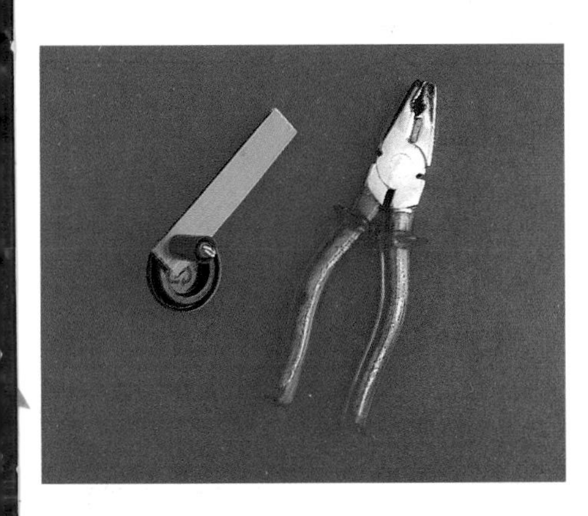

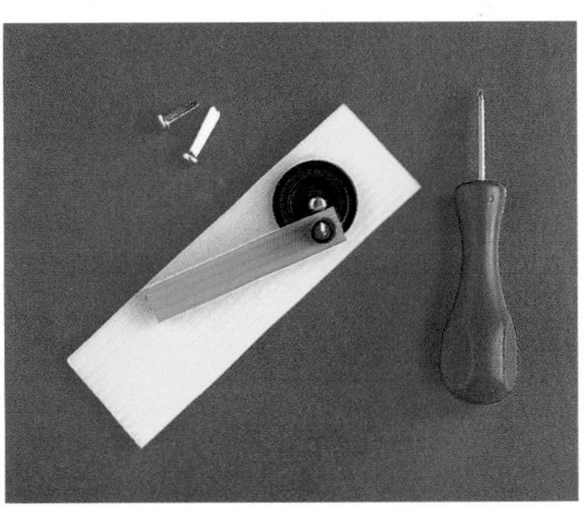

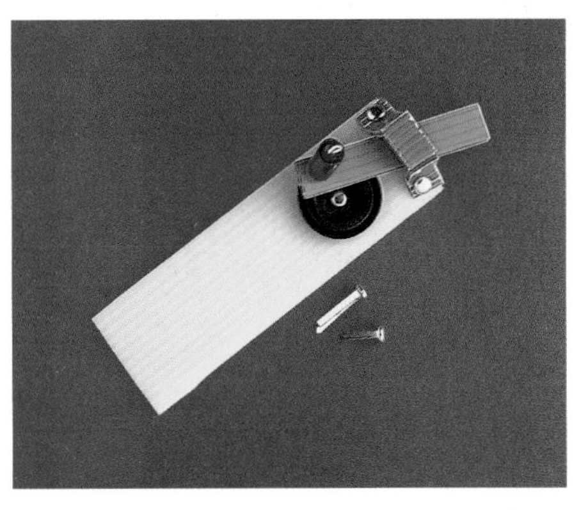

3 Turn the film canister top upside down (with the lowest part in the center). Make a hole in the side. Push the paper clip ends through the hole in the longer plastic strip, then through the side of the lid. Open them out in the groove underneath.

4 Make a hole in the center of the film canister top. Join it to the large piece of corrugated plastic with a paper fastener.

5 At the end of the large piece of corrugated plastic, make a bridge over the plastic strip with the other small strip of plastic, and fasten it to the base with two small paper fasteners.

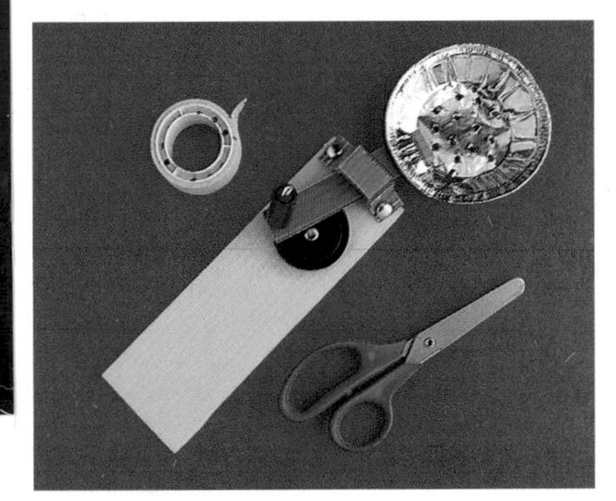

6 Make many small holes in the base of the pie pan. Cut a slot halfway down the side of the pan and push the plastic strip though it. Stick it on underneath with tape.

7 Turn the pen-top handle in a circle to spread sugar on your cereal.

BUBBLE MACHINE

YOU WILL NEED

- Two rectangular plastic boxes with lids
- Two or three other plastic containers and lids
- Medium-thickness wire, long enough to go through all the parts of the machine, plus about 4 in. (10 cm)
- Three plastic film canisters with snap-on lids
- Seven plastic beads that will fit on the wire
- Plastic packing tape
- Pliers
- Bradawl
- Scissors
- Marker pen that will draw on plastic
- Ruler
- Bubble mixture

Be careful when using a bradawl to make holes. See page 18.

This machine can work in two ways: turn the handle and make bubbles by blowing or put the machine outdoors on a windy day. If you are using the machine outdoors, you will need to put something heavy in the bottom box and fix its lid on tightly.

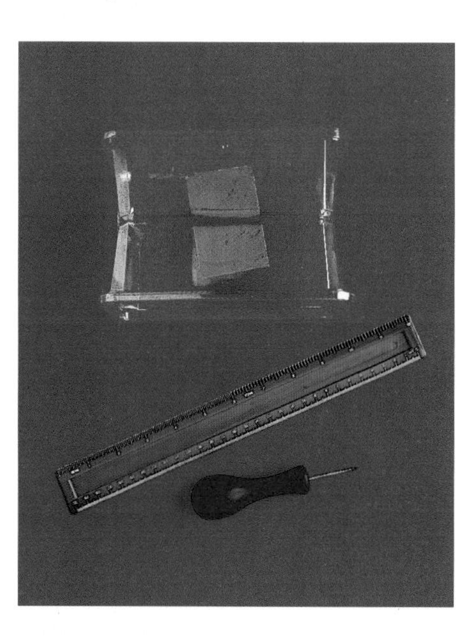

1 Measure to the center of the long sides of one box. Using the bradawl, make a small hole on either side. Put the other box underneath, upside down, and use packing tape to join them together.

Wind was used to drive machines before electricity. These Dutch windmills had sails to catch the wind. Inside the windmill was machinery for grinding grain into flour or for pumping water out of low land.

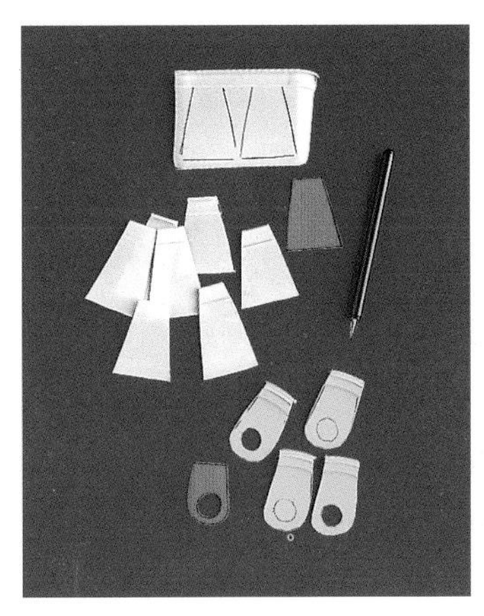

2 Using patterns (right), draw eight wind vanes and four bubble paddles. Draw them on the sides and lids of the spare plastic containers. Keep the narrow ends of the patterns on the rims. Cut them out.

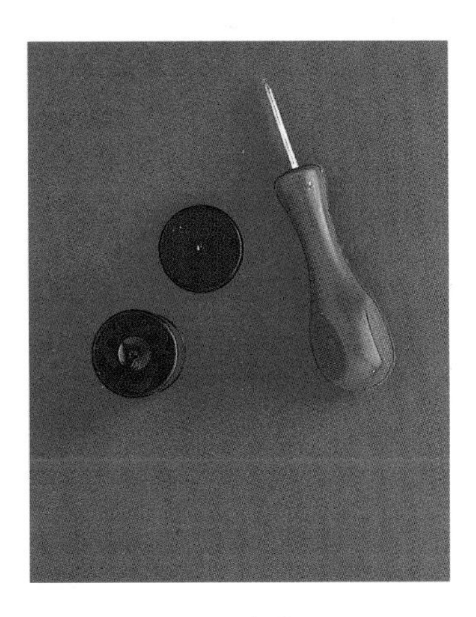

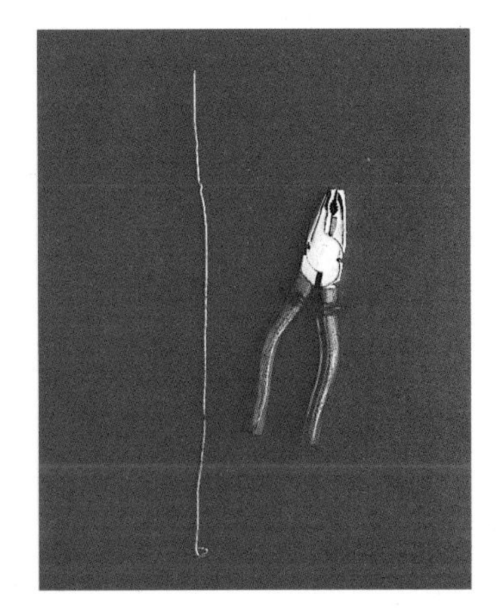

3 Carefully make a very small hole in the center of the tops and bases of the film canisters. They must fit very tightly on the wire.

4 On each film canister, cut four evenly spaced slits down the sides. They should go three-quarters of the way down. Slide the rim ends of the vanes into the slits, and put the tops on the canisters.

5 Cut a piece of wire long enough to go through all the pieces, plus about 2 in. (5 cm). Make a small loop on one end of the wire.

6 Thread the other end through a bead, a film canister, two beads, the top box, a film canister, the other side of the box, two beads, another film canister, and two more beads. Bend the end of the wire into a handle.

7 Fill the top box with bubble mixture and you are ready to go.

NOW TRY THIS

Design different shapes and sizes of wind vanes and bubble paddles. Different shapes may work differently. See what happens if you make either one bigger.

COOKIE CRUSHER

YOU WILL NEED

- Medium-size cardboard carton with top opening, about 8 x 6 x 10 in. (20 x 15 x 25 cm)
- Stiff wire
- Flat plastic or foil food tray
- Empty food carton, filled with pebbles or sand
- Five pieces of wooden dowel, about 2 in. (5 cm) longer than the width of the box
- Five thread spools
- String
- Tape
- Pliers and scissors
- Pencil
- Ruler
- Cookies or crackers

You never know when you might need to share a cookie with a friend or make some crumbs to feed the birds. Pulley power lifts a heavy weight to break a cookie or cracker.

In these photographs the back of the box has been opened so you can see what is inside. There is no need for you to do this, however, when you make the Crusher.

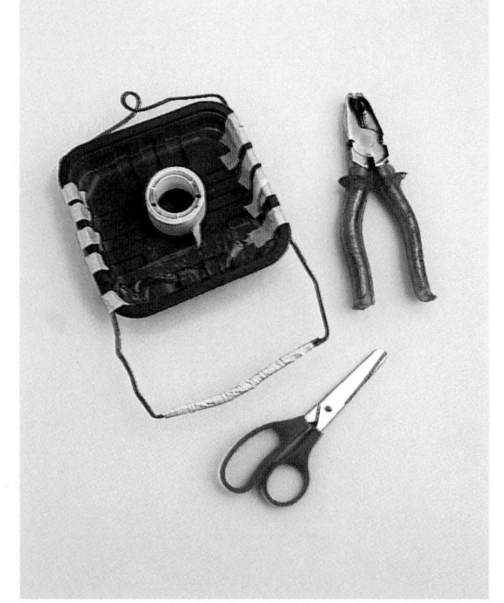

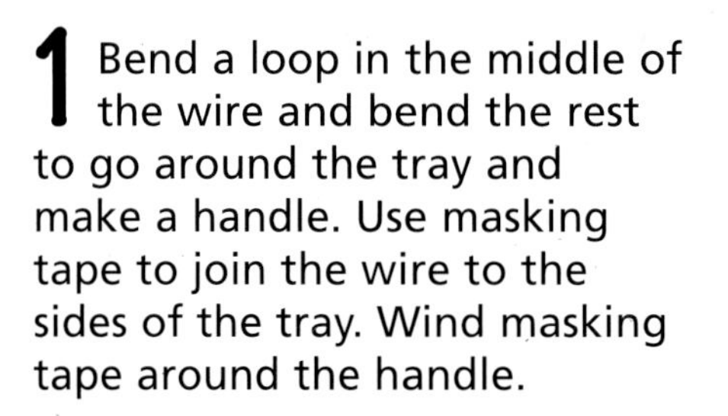

1 Bend a loop in the middle of the wire and bend the rest to go around the tray and make a handle. Use masking tape to join the wire to the sides of the tray. Wind masking tape around the handle.

2 Mark a hole in a short side of the box, big enough for the tray to slide in and out. Cut it out.

This Russian digger uses pulleys joined with cables to lift its heavy scoop filled with earth.

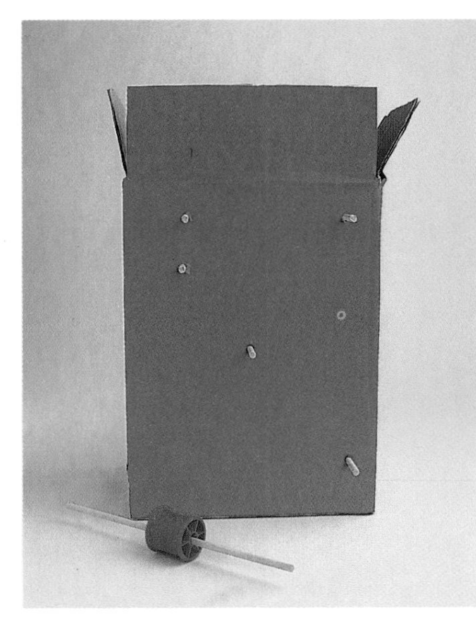

3 Make holes in the large sides of the box for the five dowel axles, in the places shown in the photograph. Push each dowel through one side of the box, through a spool, and through the other side of the box.

4 Tie one end of the string securely to the wire loop on the tray. Slide the tray into the box so the front edge is level with the side of the box. Put the weighted carton in the tray.

5 Put the string around the spool pulleys as shown on the box in the photograph. Pull it tight and and mark where it gets to the top of the weighted carton.

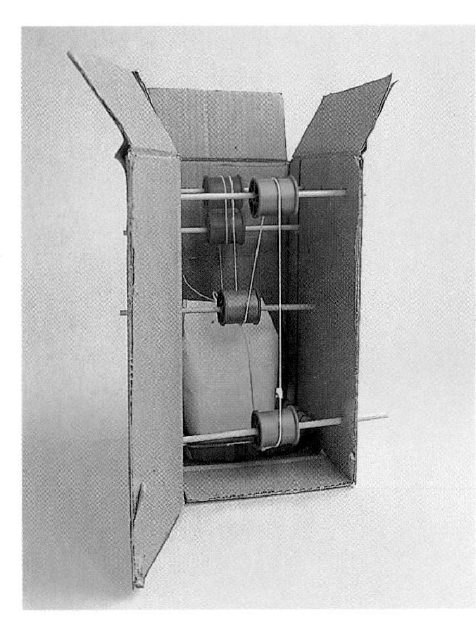

6 Take out the axles above the weighted carton. Tie the string to the carton so that the mark is exactly at the top. Put the carton, axles, and pulleys back with the string wound around the pulleys. Wind tape around the ends of the axles to keep them in place.

NOW TRY THIS

Try to make a printing machine. Put a piece of paper in the tray and stick a rubber stamp under the weighted carton.

7 Hold the whole box steady and pull out the tray. Put a cookie in the tray and let go of it quickly. The heavy carton drops down and breaks the cookie.

TWIRLER

YOU WILL NEED

- Cardboard box with lid, such as a shoe box
- Wooden dowel, .25 in. (5 mm) diameter, about 2.5 in. (6 cm) longer than the width of the box
- Wooden dowel, .25 in. (5 mm) diameter, about 1.5 in. (4 cm) long
- Seven disks cut from cardboard, about 3.5 in. (9 cm) diameter
- Matchsticks or toothpicks, cut to about 1 in. (2.5 cm) long
- Glue
- Tape
- Hole punch
- Pliers
- Bradawl
- Ruler
- Scissors

Be careful when using a bradawl to make holes. See page 18.

Cogs are used in many different places. This machine is being used to crush sugarcane to get out its sweet juice.

Cogs change a movement from one direction to another. Once you have made this machine, you can design a shape to fasten to the twirling wire. This could be a dancer or an ice skater, a model of an atom, or the moon going around the earth.

1 In one disk. Make two small holes: one in the center and another near the edge. Glue the small piece of dowel in the outer hole, then glue the disk onto one end of the long dowel.

2 Glue the other disks together in two lots of four. Let the glue dry. Make a small hole in the center of one thick disk. Mark out the top of both disks into eight equal parts.

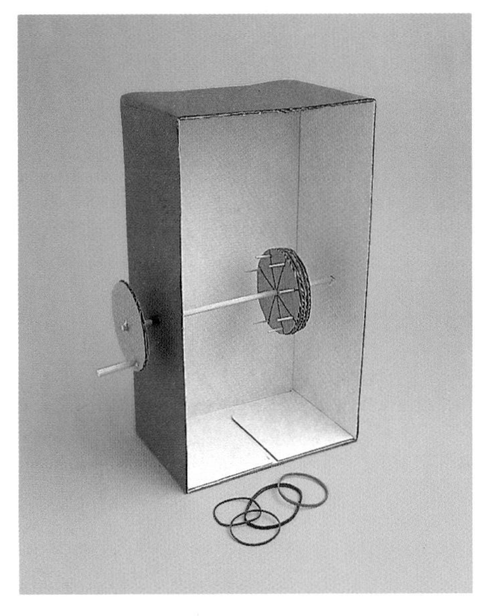

3 On each disk, use the bradawl to make eight holes on the lines, about .25 in. (6 mm) in from the edge. Glue in the matchsticks and let the glue dry. The disks make two cogs.

4 Make a hole in the same place on each side of the box, about 2 in. (5 cm) down from the top edge. Wind a rubber band around the dowel near the handle, push the dowel through one hole, through the cog, and out through the other side. Wind a rubber band around the end of the dowel.

5 Bend one end of the thin wire into a triangle base. Tape it to the back of the other cog. Cut a hole in one end of the box lid so you can see what is going on underneath.

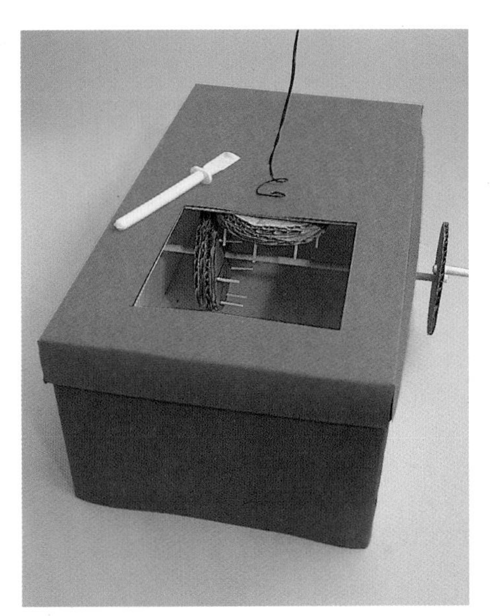

6 Make a small hole in the center of the box lid, thread the wire up through, and bend it into a flat shape to hold the cog in place underneath.

7 Slide the cog on the dowel away from the center so that the teeth of the two cogs meet when the handle is turned. Put a little glue around the hole and the dowel and let it dry.

NOW TRY THIS

Try making the sizes of the cogs different. You could also change the numbers and the lengths of teeth on the cogs to see what happens.

BELLOWS

Bellows are used to push air. They are used to make a fire burn better by blowing air onto the burning wood or coal. Old pipe organs in churches used to use bellows to make their sound.

YOU WILL NEED

- Shoe box with tight-fitting lid
- Large plastic grocery bag, with the top and bottom cut off to form a short tube
- Piece of corrugated cardboard, same width as box and 1½ times as long (corrugations should run lengthwise)
- Piece of paper 4 x 4 in. (10 x 10 cm)
- Large rubber band
- Tape
- Compass and ruler
- Pencil
- Scissors

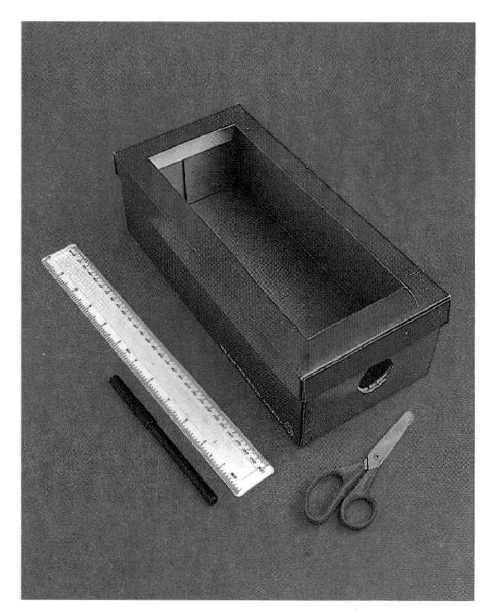

1 Mark all the way around the box lid, 1 in. (2.5 cm) in from the edge. Cut out the center. Cut a 1 in. (2.5 cm) hole in the center of one end of the box.

2 Use tape to attach one edge of the plastic-bag tube along the top edge of the lid. Stick down both long sides and one short side. Make sure it is well stuck down, especially at the corners.

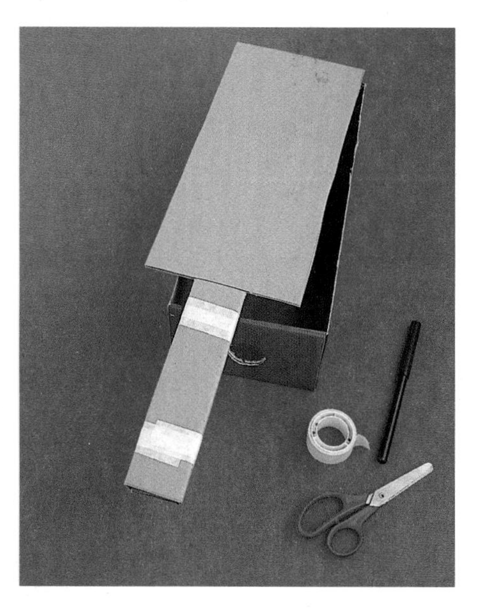

3 Put the box on the cardboard and draw a line across to mark the end of the box. Measure a third in from each long side of the cardboard. Cut and fold the sides in, to make a handle. Stick the handle together with tape.

When beekeepers open a beehive, they use a small bellows to blow smoke around the bees. This makes the bees sleepy and less likely to sting.

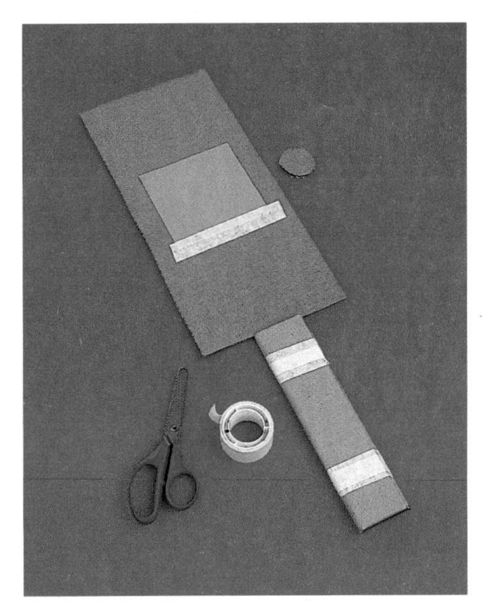

4 Mark the middle of the cardboard, and draw a 1-in. (2.5-cm) circle. Cut it out. Lay the piece of paper over the hole, and tape on end. Put the tape on the edge nearest the handle.

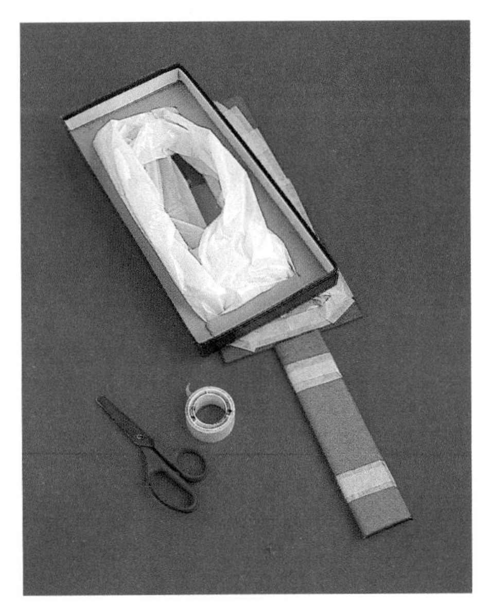

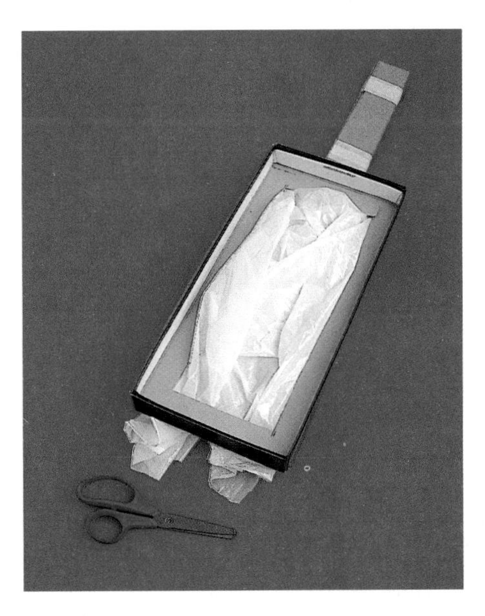

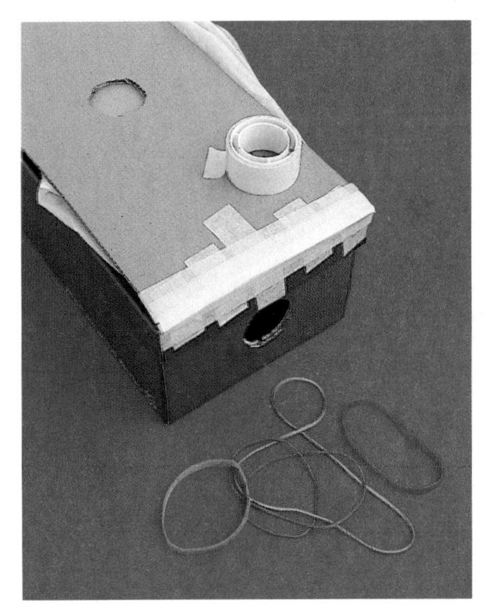

5 Put the box lid with plastic under it on top of the flap side of the handle. The untaped short end is at the opposite end from the handle. Tape the other edge of the plastic to the cardboard along two long sides and one short side.

6 Cut the bag in two at the short end, where it has not been stuck down. Fold each side and cut off the leftover plastic so that it is the same length as the box lid.

7 Turn it all over and tape across the short edge to join the handle to the box lid. Use several pieces of tape to make a strong, airtight hinge. Put the lid on the box and hold it in place with a rubber band.

FASCINATOR

YOU WILL NEED

- Strong box such as a shoe box
- Three pieces of wooden dowel, about .25 in. (6 mm) diameter: two pieces about 3 in. (8 cm) longer than width of box; one piece 1.5 in. (4 cm) long
- Two disks of corrugated cardboard, 6 in. (15 cm) diameter. Two disks 5 in. (13 cm) diameter. One disk 3 in. (8 cm) diameter
- One disk of poster board, about 10 in. (25 cm) diameter
- Thread spool
- Four .5 in. (1 cm) pieces of plastic tubing about .25 in. (5 mm) diameter, cut vertically down one side
- About 20 small rubber bands
- Glue
- Scissors
- Ruler
- Paints or felt-tip pens

This Fascinator is fun to make and fascinating to watch. Turn the handle slowly and the top disk will turn much faster. Once you have made it, you can test optical effects and illusions.

1 Make a pulley wheel from the cardboard disks. Glue together the 5-in. (13-cm) disks, then glue the 6-in. (15-cm) disks on either side. Let the glue dry. Make a small hole in the center.

2 On each side of the box, make a hole just big enough to fit the dowel so it can turn freely. The holes must be 4 in. (10 cm) from one end, and exactly opposite each other.

When you ride a bicycle, the push made by pedaling is carried by the chain to the cogs on the back wheel. This pushes the bicycle along.

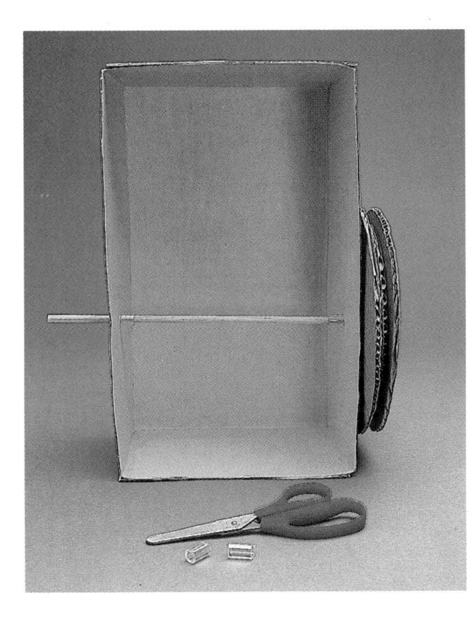

3 Push a long piece of dowel through the holes. Glue the pulley wheel onto one end. Put two pieces of tubing around the dowel inside the box so that it cannot move from side to side.

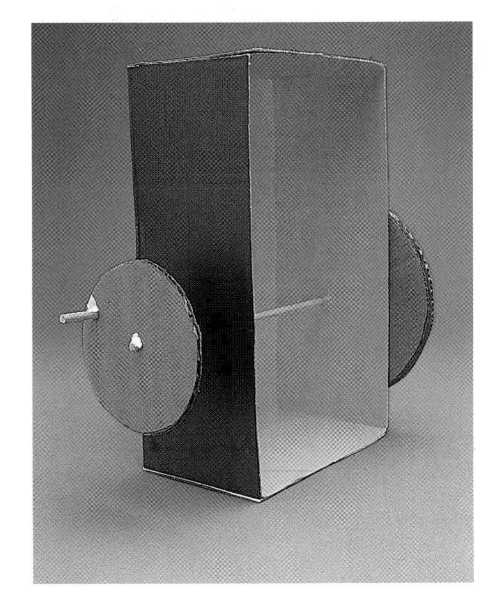

4 Make two small holes in the 3-in. (8-cm) disk: one in the center and another near the edge. Glue the small piece of dowel in the outer hole, then glue the disk on the other end of the long dowel. Let the glue dry.

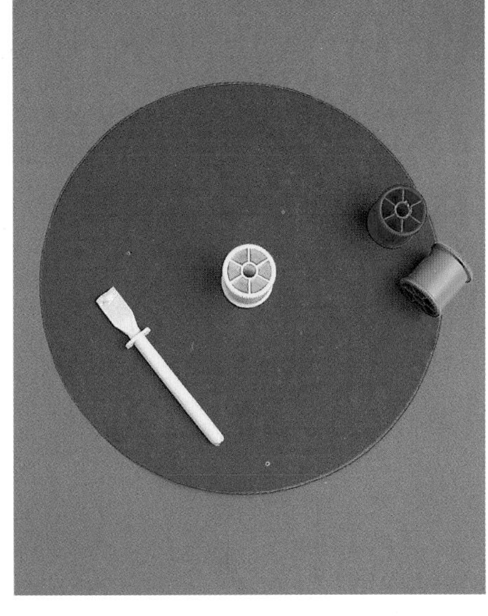

5 Draw or paint a pattern on the poster board disk. Glue the spool to the other side. Let the glue dry and make a hole in the center of the disk.

6 Make two opposite holes at the other end of the box, about 1 in. (3 cm) from the end. Push the dowel through the cardboard and spool and the holes in the box. Put pieces of tubing on the dowel at each end to hold it in place.

7 Tie together small rubber bands to make a drive belt around both pulley wheels. Tie the ends together. Turn the handle to check that the machine works. Make the belt longer or shorter if necessary.

NOW TRY THIS

Design large disks with different patterns and clip them on with paper clips. Try multicolored paper disks and see how they blend. See what happens when you use red and yellow on one disk.

GLOSSARY

axle A wood or metal rod on which a wheel turns.

cable A strong metal wire used for electricity or to pull heavy loads.

catapult A machine worked with levers and ropes to throw objects. Big catapults were used during wars to throw fireballs and rocks.

cog A wheel that has teeth sticking out from it, to turn another wheel.

cradle A framework that is specially shaped to hold an object.

crank Part of an axle or shaft that is bent at right angles to change a circular movement to a backward and forward movement.

diameter The distance across the center of a circle, from one side to the other.

drive belt A belt that goes around two different wheels, making them both move.

friction The force that slows movement when one thing rubs against another.

gravity The pull of the earth that makes things fall to the ground when they are dropped.

hinge A joint that moves, allowing objects such as doors and windows to open and close.

lever A bar that can be used to move something by pushing or pulling.

mechanism The working part of a machine.

paddle An oar or board that moves or is moved by water.

pivot A point that joins a lever to another lever or to a base that does not move.

pulley A wheel with a groove in the rim around which a rope is pulled to raise a weight or move an object.

robot A machine that works like a person.

rhythm A regular pattern of sounds, such as the sound of a horse's hooves when it is galloping.

shaft A straight pole, like an axle, that turns and carries energy via cogs or belts.

springs Curved or bent pieces of metal. A coiled spring stores energy.

template A shape used to mark and cut out a number of the same shapes.

tension The strain put on something by stretching or twisting it, such as a rubber band.

vane A flat object or blade that is moved by air or water, for example the blades of a windmill.

BOOKS TO READ

Ardley, Neil. *How Things Work* (Eyewitness Science Guides.) New York: Dorling Kindersley, 1995.

Bender, Lionel. *Invention* (Eyewitness.) New York: Knopf Books for Young Readers, 1991.

Macauley, David. *The Way Things Work*. Boston: Houghton Mifflin, 1988.

Sandler, Martin W. *Inventors* (Library of Congress Books.) New York: HarperCollins, 1996.

Snedden, Robert. *Technology in the Time of Ancient Rome*. Austin, TX: Raintree Steck-Vaughn, 1997.

Tythacott, Louise. *Musical Instruments* (Traditions Around the World.) Austin, TX: Thomson Learning, 1995.

ADDITIONAL NOTES

In presenting these projects, the authors have been very aware of the need to keep a balance between clear instructions and encouraging children to develop their own solutions.

Teachers have always used recycled materials, especially for arts and craft activities. Although store-bought components are available, all these machines have been built with scrap materials with the exception of a few items.

Waving Arms It is possible to build up long, complex systems of linked levers. It is also a good exercise to look at a machine or picture of a machine and to analyze the sequence of movements, as many are directly related to levers—a wheel can be considered as a lever rotated through 360 degrees.

Mysterious Rollers This simple toy is related to many practical applications we take for granted in modern life: escalators, audio tape and videocassettes, movies, television teleprompters, car seat belts, printing presses, and conveyor belts.

Flying Messenger A chance to experiment as to which way to wind the propeller to send the messenger in a particular direction. Have plenty of elastic bands ready as tolerance (which in itself can be tested) is strained.

Weigh-Your-Savings Bank This machine allows for the exploration of calibration and suspension. Calibrated scales can be seen but the mechanism is often hidden. The making can be linked to projects on transportation (truck weigh-stations), cooking/healthy eating, and shopping.

Drum Machine An electric motor could easily be rigged to turn this type of mechanism, or several linked versions. It could be adapted by fitting the crankshaft inside a box with wires instead of threads, to move figures on the top or sides of the box.

Mouse-A-Pult Gone are the days when children used the same leverage to catapult wads of paper around the classroom, but many such examples of the use of the lever can be noted. For less able-bodied children, these mechanisms enable extension of movement and energy.

Mini Spreader This is a chance to work at a smaller scale. Children might like to list similar mechanisms that kitchen tools use, and explore toolboxes for examples of springs, wheels, and levers, etc. It is important to use the corrugations in the plastic in the directions shown in the photographs, both for ease of bending, where necessary, and for strength.

Bubble Machine This could lead to a discussion on the properties of different materials and the reasons for choosing them.

Cookie Crusher Pulley wheels are often thought of as lifting devices; they can also be lowering and dropping systems. Many of the mechanical linkages in this book have counterparts working in the opposite direction, with energy intake and result reversed: e.g., windmill to fan, crankshaft and handle to bicycle pedals and wheel.

Twirler The axle could have two cogwheels working two horizontal wheels in opposite directions. Also, this mechanism could be linked to an exploration of methods of dividing circles by measuring angles and by drawing arcs, etc.

Bellows It may be necessary to tape all around the lid to prevent air from escaping. Early vacuum cleaners were made with two bellows, worked by each foot in turn, with valves set the other way to suck rather than blow.

Fascinator This is a very simple gear system. Experiment with different sizes and ratios of pulley wheels and compare them with bicycle systems. Two or more drive wheels could be joined, with several drive belts turning separate pulley wheels at different speeds.

INDEX

Acknowledgments

The author and publishers wish to thank the following for their kind assistance with this book: models Kitty Clark, Suhyun Haw, Yasmin Mukhida, Toby Roycroft, and Ranga Silva. Also Gabriella Casemore, Kevin Jolley, Zul Mukhida, Ruth Raudsepp, Philippa Smith, and Gus Ferguson.

For the use of their library photographs, grateful thanks are due to Chapel Studios p8 (Zul Mukhida), p13 (John Stevens), p24 (Zul Mukhida); e.t.archive p16 (Biblioteque Nationale Paris); Eye Ubiquitous p5 top (P Thompson), p11 (G Redmayne), p14 (F Leather), p27 (P Seheult); Topham Picturepoint p5 bottom.
All other photographs belong to the Wayland Picture Library: p4, p6, p19 (C. Fairclough), p20, p22, p28.